MACHINERY'S REFERENCE SERIES

EACH NUMBER IS ONE UNIT IN A COMPLETE
LIBRARY OF MACHINE DESIGN AND SHOP
PRACTICE REVISED AND REPUB-
LISHED FROM MACHINERY

NUMBER 38

GRINDING AND GRINDING MACHINES

SECOND REVISED EDITION

CONTENTS

Copyright, 1910, The Industrial Press, Publishers of MACHINERY,
49-55 Lafayette Street, New York City

WINDHAM PRESS
CLASSIC REPRINTS

In the present—the second—edition of this Reference Series Book, two additional chapters on "Types of Grinding Machines" and "Economy in Grinding" have been included. In order to provide space for this additional material, the chapters on "Lapping Flat Gages" and "The Rotary Lap," included in the first edition of this treatise, have been eliminated. These two chapters, together with additional material relating to gage making and lapping, are included in MACHINERY'S Reference Series No. 64, "Gage Making and Lapping."

CHAPTER I

TYPES OF GRINDING MACHINES

History of the Universal Grinding Machine

The universal grinding machine has had so great an influence on modern machine shop methods, and has done so much to raise the standard of workmanship and to increase the economy of production that a few words relating to the history of the development of this machine may be of interest.

The origin of the modern universal grinding machine is found in the crude grinding lathes of the early sixties. Mr. Joseph R. Brown, senior member of the firm now known as the Brown & Sharpe Mfg. Co., was intimately connected with the development of these grinding

Fig. 1. Universal Grinding Machine made by the Brown & Sharpe Mfg. Company

lathes into the universal grinding machine. The grinding lathe, as first built at the Brown & Sharpe Mfg. Co.'s works, was intended for the accurate and economical manufacture of the company's own products, and there was no idea of putting the machines on the market. In this respect the origin and development of the grinding machine was very much like the origin and development of the universal milling machine. The first work for which the grinding machines were designed was for grinding needle bars, foot bars and shafts of the Wilcox & Gibbs sewing machines. The first machine was built in 1864 and

1865, and one of these early machines is still in use in the Brown & Sharpe works. Cylindrical grinding, however, was done at the Brown & Sharpe works as early as 1862, this being indicated by the existence of drawings of a back-rest, dated September 22, 1862, which contains the essential features of a solid grinding machine back-rest of to-day.

These early machines were not grinding machines in the present meaning of the word, but were grinding lathes using, to a considerable extent, the parts of a 14-inch Putnam lathe. A great number of these were sold both in this country and abroad. Mr. Brown, however, realized the need of building a new machine designed especially for grinding, and in 1868 the design for such a machine was made. This design

Fig. 2. Plain Grinding Machine made by the Landis Tool Company

shows a machine containing most of the essential elements of the universal grinding machine of to-day. None of these machines were built at this time, however, on account of the pressure of other matters, and it was first in 1874 that working drawings were made for a complete machine containing practically all the features of the modern universal grinding machine. The first of these machines was exhibited at the Centennial Exposition in Philadelphia, 1876. The judges at the Centennial Exposition were especially impressed with this universal grinder.

The development of the grinding art since the Centennial Exposition has brought out many improvements and refinements, but the essential

Fig. 3. Heavy Grinding Machine built by the Norton Grinding Co.

principles applied to modern universal grinding machines are practically the same as those employed by Mr. Brown in these early machines.

Types of Grinding Machines

In Fig. 1 is illustrated a No. 2 universal grinding machine as built by the Brown & Sharpe Mfg. Co., Providence, R. I. The features of this machine are so well known as to hardly warrant a detailed description. It consists of a head- and tail-stock mounted on a swivel table, turning on a central stud and clamped at the ends to the main table, the travel of which is automatic. The wheel is mounted on a swivel, so that it can be set to any angle. The swivel, in turn, is mounted on a slide. The machine is provided with 'automatic cross-feed and reversing mechanism, and hence is entirely automatic in its action after the work has been once set and the machine started. Universal grinding machines are adapted for all classes of cylindrical and taper grinding.

In Fig. 2 is shown a grinding machine known as the 12-by 32-inch Landis plain grinder, built by the Landis Tool Co., Waynesboro, Pa. This machine is also of a well-known type and is distinctly a manufacturing machine. It is adapted for grinding plain straight and taper work, and all work that can be revolved on two dead centers. These machines are made especially heavy and represent the highest development of machines of their class.

In Fig. 3 is shown a grinding machine of a heavier type than the two previously shown. This machine is built by the Norton Grinding Co., Worcester, Mass., and is known as a 20- and 32- by 168-inch self-contained grinding machine, the name being derived from the fact that the machine swings 20 inches, except in the gap indicated, where it will swing 32 inches. This particular machine is motor-driven, the motor being of 20 horsepower capacity for large and heavy work. The foot-stock is arranged for sensitive adjustment in order to secure straight work. The grinding wheel used in this machine is 24 inches

Fig. 4. Internal Grinding Machine made by the Bath Grinder Company

in diameter by 4 inches face. One of the principal features of the machine is the gap arrangement, the gap being 44 inches long. The plates shown under the grinder bed are intended to be carefully embedded to straight-edge and level in the foundation of the machine. Wedges are used under the machine resting on these plates. These wedges are adjusted with two nuts, one on either side of the projecting end of the wedge, the nuts being operated on threaded studs fixed in the base of the machine. Both the plates and the wedges are machined to insure broad permanent contact. The total length of this machine is 22 feet, and it weighs 25,000 pounds.

In Fig. 4 is shown the Bath duplex internal grinding machine. When this machine was placed on the market in 1909 it represented an entirely new departure in internal grinding machines. This machine places internal grinding on as practical a basis as that which external grinding has achieved during the past decade. The principal advantage of the Bath duplex grinder is that the arrangement of the grinding spindles and the work-holding head, or heads, makes it possible to gain considerable time in the grinding and gaging of internal work. Two pieces can be ground on the machine simultaneously, and it is not necessary to shift the reciprocating slide in order to gage, insert or remove the work. It is possible to use two grinding wheels at once, one operating from each end of the work. It is also possible to use a number of grinding wheels mounted on a supported spindle between the two grinding heads and to quickly grind the inside of a sleeve or bushing by having one wheel after the other enter the work, the previous wheel, of course, leaving the work before the next one enters. This saves considerable time, as it makes it unnecessary to reverse the reciprocating table for each cut.

Fig. 5. Face Grinder made by the Diamond Machine Company

The novel feature which, in particular, distinguishes this machine from older designs is that the grinding wheels and spindles pass in through the back end of the head-stock spindle as shown in Fig. 4, instead of running into the head-stock spindle from the front.

In Fig. 5 is shown a large face grinder designed and built by the Diamond Machine Co., of Providence, R. I. This machine is designed for the general run of surface grinding operations in the ordinary machine shop. It consists, as seen from the illustration, of a reciprocating table sliding on a long bed. The table carries the work back and forth in front of the face of a large ring emery wheel. The longitudinal table movement is obtained by an open and crossed belt reversing mechanism in the back of the machine, which is connected to the table rack by heavy gearing. The mechanism used is similar to that of a planer. Machines of this type can be profitably applied to such operations as the finishing of machine columns, pipe flanges and a great number of similar pieces requiring a plane surface, but not a high degree of accuracy.

CHAPTER II

PRINCIPLES OF GRINDING*

The development of the grinding machine has made rapid progress during the last few years, and the process of grinding is more and more recognized as having both economical and technical advantages, as compared with the old methods of obtaining finish. This is especially true regarding plain cylindrical grinding, and this is due chiefly to the fact that the machines for this kind of grinding are easier to build, and in general more efficient, than machines for other kinds of grinding. It is probably true, however, that there is more misunderstanding among engineers and workmen in regard to cylindrical grinding than in the case of any of the other mechanical arts. Nearly every operator has a different theory, and each maker of grinding machines has his own method of grinding.

Relative Time Required for Finishing by Turning and Grinding

It has often been claimed, by people who have had long and thorough experience in regard to this subject, and whose testimony, therefore, must be considered as having weight, that time can be saved in finishing a cylindrical piece of work by taking a roughing cut with an ordinary cutting tool, leaving about from 0.008 to 0.010 inch of metal, and grinding off this amount instead of taking a second cut in the lathe and finishing the piece by filing. One great advantage of the grinding machine is the closer finish that can be obtained. Mr. C. H. Norton, President of the Norton Grinding Co., Worcester, Mass., in a lecture before an engineering class at Columbia University, dealt with the question of relative time and cost of finishing by grinding and turning as follows:

"A shaft 6½ inches diameter, 10 feet long, rough turned cheaply to within about 1/32 inch of the required finished size, can be finished straight, round and to size by turning with a tool that cuts off a chip removing the 1/32 inch from the diameter and leaving a good surface that requires some filing and polishing with emery cloth, and the time required to thus finish the shaft to a limit of 0.0005, plus or minus, is from seven to eight hours, according to the quality of the cutting tool and of the material in the shaft, and to the skill and ambition of the workman. To remove the 1/32 inch from the diameter, and finish the shaft to a limit of 0.0005 inch, plus or minus, by the grinding method, with a modern grinding machine, requires from one to two hours according to the ambition of the operator, and the finish is superior to that obtained by the other method. Here is a case where the grinding wheel, cutting the material to a powder with microscopic cutting

* MACHINERY, April, 1908, and May, 1909.

points, is more economical than the steel tool cutting a chip large enough to be seen and handled.

"The statement that it requires enormous power to grind steel to a powder, while cutting it into larger chips does not, should be given careful consideration. In the case of the 10-foot bar we use an average of approximately eight horsepower from one and one-half to two hours when grinding. When turning to finish we use about two horsepower from seven to eight hours. In the case of the 10-foot bar we require a nice surface and accurate measurements to a thousandth of an inch or less, and in both the lathe and the grinding machine we remove 1/32 inch, more or less, from the diameter. The production of such a grade of work when removing 1/32 inch, more or less, shows greatest economy by the grinding method. When, however, the surface can be very rough and the diameter may vary within much larger limits, a steel tool cutting deeply will remove the same amount of metal in shorter time. If the object is simply to remove a certain number of pounds of metal, turning it off with a steel tool is cheapest. But, as we know, nearly all round work must have accurate, or approximately accurate, diameter, and from approximately smooth to very smooth surface. The great majority of round work must finally have a better surface and more accurate dimensions than can be obtained with a steel tool when it is cutting at sufficient speed and depth to enable it to remove material faster than by grinding. Therefore, the limited amount of material that is removed by the finishing operation on accurate, and approximately accurate, work can be more economically removed by grinding than by turning.

"There are also cases when work not requiring a smooth surface may be more cheaply ground than turned. A case that illustrates this is that of some bridge pins, which were from 12 inches to 8 feet long and from 3 inches to 18 inches diameter. These pins were roughly turned from the billet to within 1/32 or 1/16 inch of the required diameter, and then completed by another, or sizing, cut. The limit of variation from size is 0.010 inch; if the turning tool is made to cut the right diameter at the starting end, the cutting edge of the tool must not wear off quite 0.005 inch when doing the work of removing the steel from the entire length of the pin. Now, in order to ensure that the tool will not wear away enough to cause an error beyond these limits, it becomes necessary to revolve the work so slowly that the same results can be obtained more quickly by grinding, although the surface may be far from smooth. Grinding is accomplished by a number of rapid cuts, and during the final or light cuts the grinding wheel does not wear at all, so that we are enabled to produce work of uniform diameter regardless of the length.

"While practically all round work is turned before grinding, there is a portion of such work that is most economically ground without turning. Owing to certain shapes or structural weakness it sometimes becomes difficult to turn; in such cases grinding is more economical than turning. An extreme case of this kind is that of a shaft, or bar of steel, 9/16 inch diameter and 10 feet long, with 1/16 inch to be removed from the diameter to produce an accurate ½-inch bar within a limit of 0.0005 inch, plus or minus. It is easy to understand how difficult it would be to turn this bar. It is, however, very easy to grind such a bar to the limits, and in short time. The roughing cuts that take the place of the turning easily remove the stock to within a few thousandths of an inch in about ten minutes, while hours would be consumed in turning such a bar to even coarse limits.

"In the case of slender work that springs badly when it is turned, the work can, many times, be ground more quickly than it can be turned and ground; because, when grinding off the material, the spring is ground out as it occurs, owing to the many cuts, or passes, of the grinding wheel; while when it is turned, with one cut over the piece, it must be straightened before the finishing cut is taken. It is true, however, that the majority of work should be turned before grinding."

Another fact in connection with cylindrical grinding to which Mr. Norton calls attention is that in order to secure the greatest economy by the use of grinding machines, less should be paid for the turning than when the work is to be finished in the lathe. With well constructed grinding machines, the coarser the turning, the quicker the grinding can be done. It is no longer necessary to turn either smoothly or correctly to size. A variation of 1/32 inch, more or less, on large work is of no moment, and on small work a variation of 1/64 inch is permissible, and the surface may be very rough in all cases. A large part of the economy is secured by cheap turning.

Grinding Hard Alloy Steels

In some special cases, when the steel to be finished is so hard that it cannot be cut by means of a cutting tool, the grinding machine has to take the place of the lathe entirely. Of course, the work in this case cannot be done so cheaply as in the case of ordinary kinds of steel, but still it can be done with fair economy. As the piece is taken entirely rough and put up in the grinding machine, there is, considering the errors in casting. about 1/8 inch up to 3/16 inch on the diameter that has to be ground off. When so large an amount of metal has to be removed by grinding, another problem than that dealt with when only 0.008 to 0.010 inch has to be ground off, presents itself. The writer at one time designed three special grinding machines, two for external and one for internal work, all for very heavy duty. Herein are given a few of the conclusions arrived at while designing these machines. Being, as mentioned, mostly used for heavy grinding, the machines may differ some from the common light grinding machines, but the principles remain, in general, the same.

Any machine tool must, of course, be designed heavy enough not only to take all the strains produced by the action of the cutting tool or wheel, but to prevent all, or nearly all, vibration and chattering of the machine itself. This is true of the grinding machine more than of any other machine tool. Rigidity is a very important factor in the efficiency of the machine, both in regard to heavy grinding and grinding for very exact sizes and high finish.

Influence of Vibrations on Action of Grinding Machines

The grinding wheel rotating at a high speed tends to jar its bearings and supports. Vibrations of this kind would result in an oscillating motion of the grinding wheel perpendicular to its own axis of rotation and along the line connecting the center of work with the center of the wheel. The frequency of these vibrations depends entirely upon the weight of the oscillating parts. The cause of the vibrations is that the center of gravity of the rotating parts, grinding wheel, shaft, pulley, etc., is not entirely the same as the center line of rotation. This is partly due to the uneven structure of the material. It is very plain to everybody that the oscillating grinding wheel cannot cut to its full capacity. The length of the oscillations might not be-

TABLE I. SPEED OF EMERY WHEELS

Diameter of Wheel, inches	R. P. M. for Surface Speed of 5,000 feet	R. P. M. for Surface Speed of 6,000 feet
1	19,099	22,918
2	9,549	11,459
4	4,775	5,730
6	3,183	3,820
8	2,387	2,865
10	1,910	2,292
12	1,592	1,910
14	1,364	1,637
16	1,194	1,432
18	1,061	1,273
20	955	1,146
24	796	955
30	637	764

large, perhaps only one-thousandth of an inch or a fraction thereof, but the cut will be just so much deeper one moment than the next following. Only at one moment, when the wheel is furthest in, will it cut to its full capacity.

It is very important, in order to secure nice running of the wheel, to have the belts in good order, and to have the boxes closely adjusted, even though they run a trifle warm. Because of the high speed of the shaft, the boxes ought to be made with ring-oiling devices. This would allow a closer adjustment, and secure a better running of the shaft. However, as far as the writer knows, there are no grinding machines on the market equipped with ring-oiling boxes. The slides should, for the same reason as the boxes, be adjusted closely, even though they slide hard.

Speed of Grinding Wheels

The peripheral speed of the grinding wheel should be approximately from 5,000 to 5,500 feet per minute. There are occasionally cases when higher speed is desirable, but with higher speed there is danger of the wheel breaking. The wheel should, however, never be run slower than 5,000 feet per minute, because it becomes less efficient at slower speeds.

Above will be found a table which gives the number of revolutions per minute for specified diameters of wheels to cause them to run at the respective periphery rates of 5,000 and 6,000 feet per minute.

Experience has shown that for grinding work with fairly large diameter, better results are obtained by using a comparatively small wheel than by using one with too large a diameter. The explanation of this fact is that the wheel of smaller diameter clears itself faster from the work, while the larger one has a larger contact surface, and, therefore, the specific pressure between wheel and work becomes reduced, and the metal removed by the wheel stays too long a time between the wheel and work, and prevents the particles of the wheel from cutting properly into the work. The peripheral speed must, however, be the same for the smaller wheel as for the larger one.

Surface Speed of Work

The proper surface speed of the work varies somewhat with the material and kind of work to be done. The grinding machine builders recommend 15 to 30 feet as a good average speed range for ordinary kind of work. For cast iron this can be slightly increased. The writer has had experience in grinding a very tough and hard steel (manganese steel), and has found the right surface speed in this special case to be as low as 6 to 8 feet a minute for rough grinding. For the finishing grinding, the speed should be somewhat higher than for the rough grinding. For delicate work the speed should be slow, because the work could easily be damaged by forced grinding.

As a general rule, for determining the surface speed for a certain kind of material, one can say that a brittle material, as cast iron, takes a high speed, while a tough and hard material, as the best tool steel, takes a slow speed. For grinding close to size and for high finish, the depth of the cut must be small, and higher surface speed can consequently be used.

Many of the grinding machines on the market are built so as to have the work revolving on two dead centers. This is done more for the sake of being able to obtain accuracy than for the sake of increasing the cutting efficiency of the machine.

Traverse Speed of Grinding Wheel

The traverse speed of the grinding wheel should for ordinary grinding be three-fourths of the width of the wheel, that is, for one revolution of the work the wheel should travel three-fourths of the width of the face. If the wheel be traversed slower, the new cut is overlapping the old one more than necessary, and too large a part of the wheel is idle. It is, however, necessary that the new cut overlap the old one with about one-fourth of the width of the face, because the edges of the face easily become rounded off, and, if the travel be too rapid, the result is an uneven surface.

The capacity of the wheel, within certain limits, of course, is proportional to the width of the face. A certain specific pressure between wheel and work is required for the highest cutting capacity. A wider wheel requires consequently a larger total pressure. But many of the machines now on the market are not rigid nor heavy enough to stand the pressure needed for a fairly wide wheel, cutting at full load, without vibration and chatter. The grinding machines on the market have not, in the writer's opinion, yet reached their full capacity. Wider wheels should be used, and the machines should be designed and built heavier in order to take the load of the cutting wheel, without perceptible vibration of the machine.

For the final smooth finish, a slower traverse speed should be used, especially if the face of the wheel is not kept a perfectly straight line. A smoother surface is obtained by using a slower traverse speed. The part of the wheel which is overlapping, while theoretically it does not cut, still wears away the unevenness left from the first cut, and thus to some extent polishes the surfaces.

While grinding a plain cylindrical piece of work, the grinding wheel should not be allowed to travel too far past the ends of the piece before reversing; it is only waste of time. The wheel should be reversed when three-fourths of its width is past the end of the work.

Depth of Cut

The depth of the cut to be taken depends upon the material, kind of wheel, and the work done. It should be deep enough to permit the wheel to do its utmost. This is, of course, true only about pieces that are rigid enough to stand a heavy cut. The grinding operator himself will have to determine the depth of cut for each individual case, judging it by the prevailing conditions of work, machine, and wheel.

When the piece to be ground, owing to the hardness of the material, cannot be roughly finished by a cutting tool before being placed in the grinding machine for the final finish, there is often up to 3/16 inch on the diameter to be removed by grinding. Employing the same principle as when the piece is roughly turned in a lathe, previous to the grinding operation, the work should first be put up in a machine equipped with a coarse and wide grinding wheel. A wheel of this kind is capable of removing stock rapidly. The piece should be finished to within 0.005 inch of the finished diameter in this machine, and then moved to a machine equipped with a finer grain wheel, and the final finish given to it.

The Grinding Wheel

For heavy grinding, the alundum wheel is the best for removing stock rapidly. The carborundum wheel will give a smoother finish, and is to be recommended for the large majority of other classes of grinding. Emery is less abrasive, but gives a higher polish. Most grinding wheel manufacturers recommend their medium grade, M.

The question as to what is the very best wheel for finishing any particular piece cannot be definitely answered. On the next page is given a table of wheels which can, with advantage, be used in the cases mentioned. This table is recommended by one of the largest grinding machine manufacturers.

Grit No. 24 may be too coarse for any but rough classes of work, but if mixed with No. 36 it gives a fair result. No. 30 used separately is capable of a very fair commercial finish, but if mixed with No. 46 will give as fine a finish as is desired by the majority of the grinding machine users, and at the same time it retains the rapid cutting capacity. Nos. 46 and 60 are as fine as is necessary for almost any manufacture, although finer than these are used by some concerns who require a very high gloss finish.

A satisfactory grinding wheel is an important factor in the production of good work. In machine grinding, it is desirable, in order that the cut may be constant, and give the least possible pressure and heat, to break away the particles of the wheel after they have become dulled by the act of grinding. It is the capacity of yielding to or resisting the breaking out of the particles which is called grade. The wheel from which the particles can be easily broken out is called soft, and the one that retains its particles longer is called hard. It is evident that the longer the particles are retained, the duller they will become, and the more pressure will be required to make the wheel cut. Retaining the particles too long causes what is familiarly known as glazing. A wheel should cut with the least possible pressure, and must therefore be sharp. This sharpness is maintained by the breaking out of particles. Therefore, a wheel of proper grade, cutting at a given speed of the work, possesses "sizing power," or ability to reduce its size uniformly without breaking away its own particles too rapidly;

TABLE II. GRADE OF WHEEL TO USE FOR DIFFERENT MATERIALS

Material	Grit No.	Grade
Soft Steel { Ordinary shafts ..	24 to 60	Medium.
Steel tubing or very light shafts.......	24 to 60	Two or three grades softer than medium.
Tool steel or Cast iron ....·.	24 to 60	Medium or one grade softer.
Internal grinding	30 to 36	Medium or several grades softer.

obviously if the work is revolved at a higher speed, the particles will be torn away too fast, and the wheel will lose its sizing power.

The properties of toughness and hardness of the material to be ground have a retarding influence on the grinding because they make the material stick to or clog the wheel. The ground-off material, instead of being thrown away from the wheel by the centrifugal force, gets in between the particles of the grinding wheel. It is self-evident that this has a greatly retarding effect on the cutting quality of the wheel. A brittle material, on the contrary, does not have the tendency to clog the wheel, but the stock ground off is immediately thrown away from the wheel, leaving the particles free to cut without the retarding action of undue friction, and the generation of more than the due amount of heat. If we take into consideration only these properties of the material to be ground, the tough or leady material requires a soft wheel, because the particles must break away fast enough to prevent the wheel from being clogged. In this case, the particles do not wear enough to become dull, but must break away before this. When grinding a brittle or hard material, on the contrary, the wheel is less liable to be clogged, the particles do not need to break away so soon, and, therefore, a harder wheel should be used.

However, the wheel must not be so hard that the particles get too dull and become inefficient as cutting agents before they break away.

Importance of Wheel Running True

In order to obtain the full efficiency of the grinding wheel, it must be run perfectly true; that is, cut evenly all the way around. The grinding wheel detects its own errors. A slight difference in the sparks indicates that the wheel is out of true. The eccentric wheel has about the same kind of action as the one which is vibrating because of too weak supports. Furthermore, the edge of the grinding wheel should be kept perfectly straight. If the edge be curved, however slightly, a curved cut will be the consequence. Many grinding machines

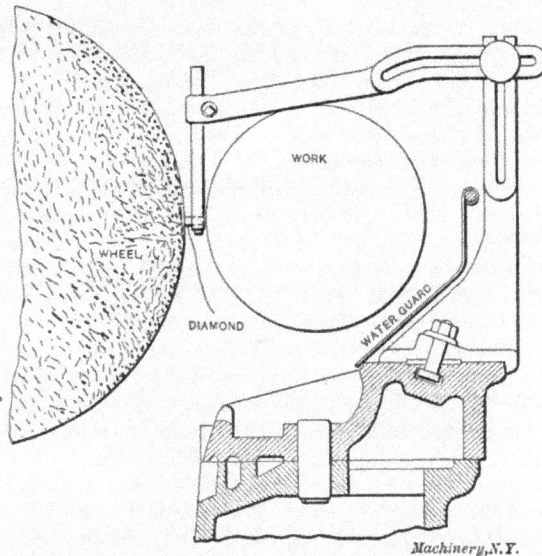

Machinery,N.Y.

Fig. 6. Fixture for Truing Emery Wheel with Diamond

give inefficient results because the edge of the wheel is not kept in a true straight line. The operator seldom appreciates the great importance of this, and, therefore, the foreman should watch the men closely in regard to this point.

The best tool for truing the wheel is the diamond, but, this being rather expensive for shops where not very much grinding is done, the usual emery wheel dresser can be used to good advantage. In truing the wheel, the dressing tool should be kept stationary and rigidly supported, and the wheel should be traversed back and forth, until a true edge is obtained. Fig. 6 shows a fixture and arrangement for wheel truing with a diamond.

Wet and Dry External Grinding

Nearly all plain cylindrical grinding is now done wet. There are many reasons why the wet method is to be preferred to the dry. Because of the friction between the grinding wheel particles and the work, as well as between the cut-off material before it leaves the wheel and the work, more or less heat is generated. If this heat is not carried away, the work will be burned. Besides, the edge of the grinding wheel would be highly heated, but the center would still remain comparatively cool, and the outside would expand and there would be danger of the wheel breaking. It is found that the water has a softening effect upon the wheel; therefore, a harder wheel is required for wet grinding than for dry.

Machines with Two Grinding Wheels

The grinding machines on the market are equipped with only one grinding wheel, but there is no reason why two grinding wheels cannot be employed to advantage. In this case one wheel is to operate on each side of the work. As both of the wheels are to throw the sparks and the water down, one of the wheels has to cut with the revolving of the work, that is, the peripheries of the wheel and the work are going downwards. This is, of course, not the ideal condition, but, when the work is revolving at a slow peripheral speed, there is not much difference in the cutting capacity of the two wheels.

It is self-evident that, when employing two wheels, one at each side of the work and just opposite each other, the traverse speed of the wheels must be twice as fast as in the case of only one wheel, or three-fourths of the width of the wheel for one-half revolution of the work. Otherwise one wheel will overlap the cut of the other.

The two machines for external grinding which the writer designed have two wheels working according to the principle previously described. Fig. 7 gives an idea of the arrangement used on one of these machines. The principal features of the design can be studied direct from the illustration without any further comments.

One new feature of these machines is that each grinding wheel is driven independently by a motor. This motor is mounted above the wheel spindle, and is belted directly to it. Special attention has been paid to designing the support of the motor in order to prevent the vibrations of the motor from being transferred to the grinding wheel.

Internal Grinding

The development of internal grinding machines has not advanced as fast as that of machines for external grinding. It has even gone so far that one man holding a prominent position with one of the largest grinding machine manufacturing concerns in the country has said that in his opinion, the internal grinding machine is a mistake from start to finish, and that it will never be made a success. This, however, is rather too broad a statement in face of some recent developments in this direction.

The statement just quoted, nevertheless, was not made without good reason. As we have already seen, the rigidity of the arrangement for supporting the grinding wheel is a very important factor for all efficient grinding. The internal grinding machine does not very well lend itself to the employment of any rigid and heavy fixtures, and the grinding wheel must necessarily be small, and therefore lacks the strength to stand a heavy cut. The designer, when designing the fixtures for internal grinding, has an entirely different problem to solve than when designing those for external grinding, where it is comparatively easy to obtain ample rigidity. The internal grinding wheel

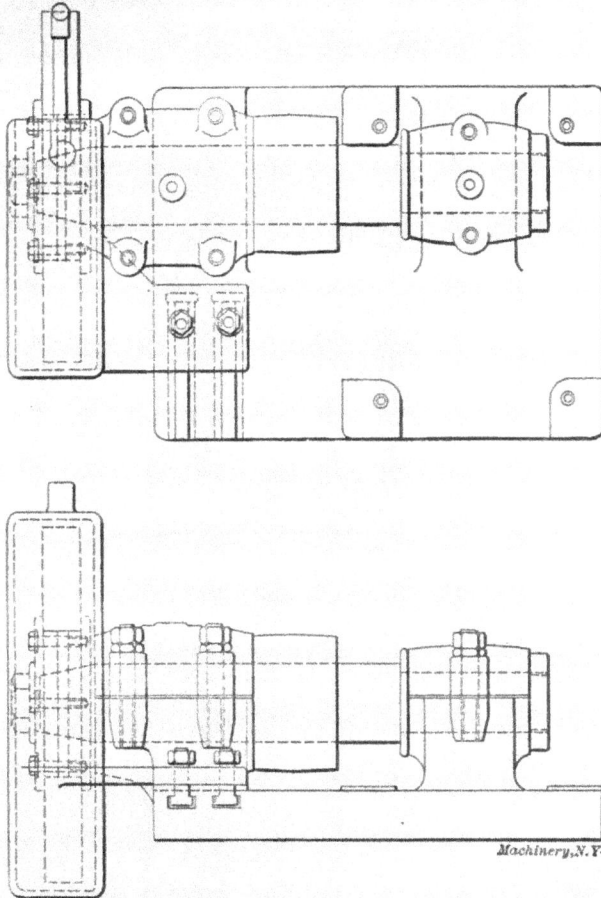

Fig. 7. Grinding Head for External Grinding Machine

must be mounted at the end of a small spindle which projects past the bearing far enough to enable the wheel to reach past the end of the hole to be ground. Such a spindle rotating at a high speed is liable to vibrate, especially if pressure be applied at the end of it, as is here the case.

Sometimes, however, it becomes absolutely necessary to grind, internally, even a comparatively large amount of stock. This is the case when finishing manganese steel, this material being so hard that it cannot be cut by any kind of tool steel. Take the case of bores of manganese steel car wheels. As the grinding of the bores must be done without any stock having previously been removed from the rough casting, on the average about one-eighth inch of metal must be ground off from the hole. All the errors in the cored hole, as eccentricity in reference to the circumference of the wheel, etc., must be corrected by grinding. A hole cored in a manganese steel casting is always comparatively much rougher than a hole cored in cast iron, and all this must be taken into consideration, when determining the amount of stock to leave for the grinding process.

Design of Heads for Internal Grinding

The fixture used in the internal grinding machine designed for grinding these wheels is shown in Fig. 8. Internal grinding fixtures generally have a long extension bearing, as shown in Fig. 9. This serves to support the spindle as near to the grinding wheel as possible; but

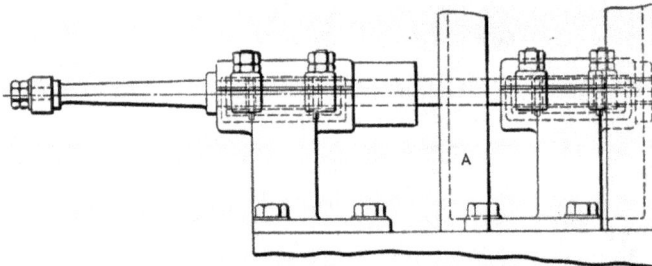

Machinery,N.Y.

Fig. 8. Grinding Head for Internal Grinding

the diameter at the root of this extension, that is, nearest to the box, cannot exceed the diameter of the grinding wheel.

The spindle, shown in Fig. 8, is made solid, and has the largest diameter possible for the size of the grinding wheel. An increased amount of rigidity and a greatly increased simplicity is gained by this design.

When working, the grinding wheel produces, especially in dry grinding, very much dust. When inside a hole the dust cannot very easily get away, but whirls about in the hole. If the spindle has a bearing near to the grinding wheel, the dust will find its way into the journal. This drawback is entirely eliminated by having a large solid spindle without a bearing near to the grinding wheel.

As to the relation between the overhanging part of the spindle and the distance between centers of the boxes, there are many factors that come into consideration in regard to this relation, such as the design of the boxes, the diameter of the spindle, how close the spindle can be allowed to run in the boxes, etc. However, the distance between the centers of the boxes should be made as large as the general design conveniently permits.

Fig. 8 shows at *A* the support for the motor. This support is placed on the top of the top rest. The driving pulley is placed between the bearings, so that the support could be made as rigid as possible.

It was found by actual experience with these fixtures, that when the grinding wheel was taking a fairly heavy cut, the spindle did not vibrate nearly so much as when the wheel was running idle. The springing quality of the spindle, and the pressure between work and wheel made the wheel cut without any chattering worth mentioning.

Regarding the peripheral speed of the grinding wheel, what has already been said with reference to external grinding is equally applicable to internal grinding.

Because of the lighter fixtures, the speed of the work should be slower than for external grinding. The writer has found the right cutting speed for hard and tough steel to be, for heavy grinding, about seven feet a minute. For the finishing, the speed can, with advantage, be somewhat higher. The wheel should travel three-fourths of its width for one revolution of the work, the same as for external grinding.

Wet and Dry Internal Grinding

One point that has been much discussed in regard to internal grinding is whether it shall be conducted wet or dry. Some grinding

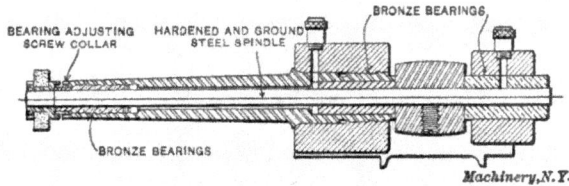

Fig. 9. Common Construction of Grinding Heads for Internal Grinding

machine designers have advanced the opinion that it, by all means, must be done dry, but others claim the wet method to be superior. For light finishing grinding one method might be considered as good as the other, because so small an amount of heat is generated that there is no danger of burning the material or breaking the wheel. But, for heavier grinding, a considerable amount of heat is generated, and it becomes necessary to carry it off by water. At least, such is the writer's own experience on this subject. At a test recently conducted to find out the actual difference between dry and wet internal grinding, it was found that the cutting quality of the grinding wheel was about the same in both cases, but, with a heavy feed and dry grinding, the work was highly heated, and the wheel broke after about half an hour's run, while, with wet grinding, the wheel stood the heavy cut continuously without breaking.

The water can be injected into the hole in a stream about 1/16 inch in diameter. In addition to carrying away the heat, the water serves to wash away the removed stock from the hole.

Tests have been undertaken on the above mentioned internal grinding machines, in order to find out the time required to grind the bores of a certain kind of manganese steel car wheels. Two different kinds of wheels were tested. The first one, a 20-inch diameter wheel, had a bore $2\frac{7}{8}$ inches in diameter and $5\frac{7}{8}$ inches long, and it was to be ground for a press fit. The second one, an 18-inch diameter wheel, had a bore $3\frac{1}{4}$ inches in diameter and $4\frac{1}{2}$ inches long, and was also to be ground for a press fit. Four wheels of each kind were ground during the course of the test, and it was found that the actual time for the grinding operation, not including the time required for putting up the work in the machine, was, for the first kind of wheels, 1 hour and 23 minutes for all four, and for the second kind, 1 hour and 9 minutes for four wheels. Considering that the bores of the wheels were not previously turned, but entirely rough, as the wheels were taken directly from the foundry, and considering the hardness and toughness of the steel, the results obtained were considered good. The time of putting the work in the machine was about 6 to 8 minutes for each wheel. As the machines work automatically, one man is able to run three machines. Counting 8 minutes for the putting up of each wheel, the man is able to grind one wheel of the first kind in 30 minutes, and one wheel of the second kind in 26 minutes.

The work was revolved at a speed of 7.7 revolutions per minute. This makes a peripheral speed, for the first case, of 5.8 feet per minute, and for the second case, of 6.6 feet per minute. The grinding wheel used was a 2-inch diameter, 1-inch face, No. 46 grit, O grade alundum wheel. It was run at a speed of 4,750 feet per minute.

The traverse speed of the work was as high as 0.84 inch per revolution of work. This allowed the wheel to overlap the old cut by only 0.16 inch, but, as the grinding wheel was trued very carefully, this was found to be all that was required for obtaining a nice smooth surface. The traverse feed was not slowed down, but remained the same while doing the final finishing, and a very satisfactory finished hole was obtained. The test was made throughout with wet grinding.

For heavy cylindrical grinding, which has especially been referred to, the width of the wheel used varies between $1\frac{1}{2}$ and $2\frac{1}{2}$ inches, regardless of the diameter. In some special cases narrower wheels than $1\frac{1}{2}$ inch are used, but these special cases are exceptions to the general practice, and must be recognized as such by the machine builders and users. Although larger wheels are used, there is no doubt that the best range of diameters of wheels is between 12 and 18 inches. For how wide a wheel the grinding machine of the future can be designed, has yet to be decided; but, wider wheels and heavier machines point the direction of the road which the designer and machine builder should follow for the development of the grinding machine.

CHAPTER III

ECONOMY IN GRINDING*

It is very often the case that a grinding machine falls short of its highest possible output by reason of the inattention of the operator to some of the short cuts and time-saving methods that have been highly developed in the use of other machine tools. It is the case with grinding machines as with other machine tools, that the development of short cuts and kinks of various sorts greatly increases the aggregate output of the machines. The lack of such time-saving methods is the reason for the unfavorable attitude of some firms to the grinding process. It is almost always the experience of a demonstrator sent out by the manufacturer of grinding machines that his results are not maintained by the operator. First one little kink is lost sight of, then another, and the time increases very slightly on each individual piece ground; consequently the aggregate of the day's output is soon considerably below what it should be.

A grinding machine will size work to a commercial degree of accuracy with remarkably little attention on the part of the operator, but the quantity output of a machine is very largely dependent on the ability and willingness of the operator to hustle; hence the reason for the almost universal prevalence of the "piece system" in the grinding department.

Spotting Work for the Back-rests

Spotting work for the back-rests is a great help to the operator in several ways. When a piece of work is placed in the machine, assuming that the automatic cross-feed stop shield shown at A, Fig. 10, has been adjusted from a previous piece ground, the grinding wheel should be run back from the work about 1/32 of an inch. This is accomplished by turning the cross-feed handwheel B about one revolution in the opposite direction to that in which it is automatically revolved. The table carrying the work can now be moved by the handwheel marked C, provided for the purpose, bringing the various back-rests successively in front of the grinding wheel, as shown in Fig. 11. The wheel is then fed into the work by hand without reciprocating the table; it should be fed into the work until the diameter where the shoes of the steady-rests bear is within one-thousandth of an inch of the finished size. The guard A, Fig. 10, as it approaches the pawl D, serves as a gage to determine the extent to feed the wheel in. This operation takes a very short time and provides a smooth surface for the bronze shoes. This surface is so near the finished diameter that the shoes are not worn large before the work is reduced to the finished size, therefore the work is accurately and steadily supported

---

* MACHINERY, May, 1910.

during the finishing cuts. This is found particularly advantageous on hardened work where a large amount of stock is left for finishing and where chatter marks are more apparent if the supporting shoes of the back-rest do not fit the work very closely. It is also found to be a great saving on the wear of the shoes themselves.

Grinding to a Shoulder

A time-consuming error among grinder operators is made on account of the prevalence of the idea that the reversing dogs on a grinding machine should be set to reverse the table traverse when the grinding wheel is within a very few thousandths of an inch of a shoulder on

Fig. 10.　Controlling Mechanism of the Brown & Sharpe Grinder

the work as at E, Fig. 11. It is a fact that most grinding machines will reverse within one or two thousandths of an inch of the same place each time, provided the rate of table travel is not changed. The variation in the depth of center holes in the work makes it necessary for the operator to try the reversing of the machine by hand after placing each piece in position to make sure that the wheel will not gouge the shoulder, as it would surely do if the center holes were a little smaller than in the piece previously ground and a close limit for reversing were used. There are two ways to prevent the wheel gouging the shoulder on the work; one is the adjusting of the reversing dog by the screw F, Fig. 10, for each piece ground, which is a time-consuming operation; the other, and better method, is to bring the shoulder on the work up to the wheel by hand (using hand-wheel *C* after spotting for the back-rests as described), then feed the wheel straight into the work, reducing the diameter next to the shoulder for a distance equal, of course, to the width of the wheel, to the finished size. This is quickly and easily done by using the knock-off shield *A* against the pawl *D* for a gage as described. The table may then reverse for the subsequent complete grinding of the piece when the wheel has advanced not closer than ⅛ inch or more from the shoulder, and the edge of the wheel next to the shoulder does not become worn away or rounded because it runs off the work, or nearly off, at each end of the piece. A wasteful truing off of the wheel is thus easily avoided.

This operation of necking the work at a shoulder with the full width of the wheel, obviates the necessity of a dwell of the table at that reversing point. If a machine dwells when reversing at the shoulder end of the traverse, it must of necessity dwell at the other end where the wheel usually runs nearly off the work; here the dwell is not only of no value but it is very likely to cause the wheel to grind the end of the work undersize. While this dwell is only momentary, it is quite a factor in a day's output that can be readily eliminated.

If for any reason the necking of the work with the wheel is not deemed expedient and a dwell is required, this dwell is needful only once or twice during the grinding of a given piece and can be produced at will by the operator pressing the knob *H*, Fig. 10. Pressing this knob stops the power traverse of the table, which may then be fed over by hand, using hand-wheel *C*, to face up the shoulder, and the dwell may be prolonged to allow one or more revolutions of the work as the particular quality of work and wheel may require, instead of the length of dwell being dependent entirely on the speed of the reciprocating table, as is the case when the dwell is automatically supplied by the gearing of the table. The table traverse is started after the dwell by pulling the knob *h* to the position shown in Fig. 10. A dwell so produced is not duplicated at the other end of the work except at the will of the operator.

It should be clearly understood that a grinding machine can be reversed with the shoulder on a piece of work ⅛ inch or more from a grinding wheel, then stopped at the reversing point and "forced over" this ⅛ inch or more beyond the normal reversing point without disturbing the reversing dogs and without subjecting the reversing mechanism to any strain.

In Fig. 13 are shown the elements of the reversing mechanism of the Brown & Sharpe plain grinders, which shows quite clearly how this traverse beyond the reversing point is accomplished. When the reversing dog *J* or *K*, Fig. 10, strikes and reciprocates the reversing lever *L*, the motion is transferred by its fulcrum stud and lever *M* to the arm *N* which compresses the reversing spring *O*. When this arm, which compresses the spring, has moved far enough to give considerable tension to the spring, the taper lug on the arm *N* raises

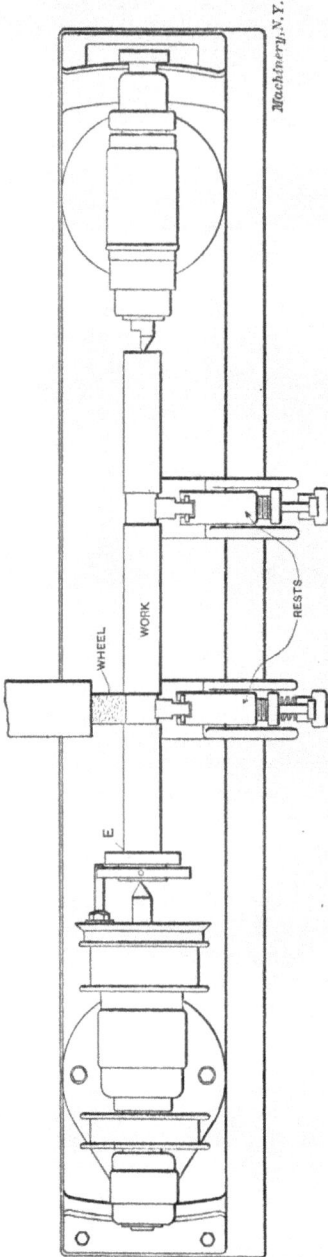

Fig. 11. Plan showing Method of Spotting Work for Back-rests

Machinery,N.Y.

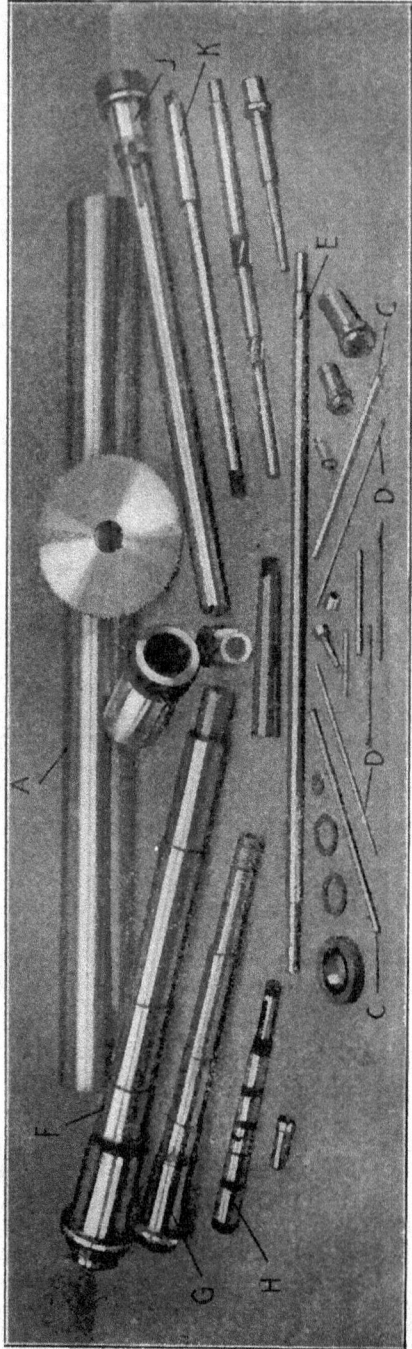

Fig. 12. Examples of Ground Work

the latch *P*. thus releasing the yoke *R* which is connected to the reversing clutch *S*, this deriving its power from gear *T*. The spring, which is under compression, throws the reversing clutch as soon as the latch releases the yoke *R*. This movement of the yoke is sufficient to relieve the compressed reversing spring so that the reversing lever can traverse farther than the position where reversing takes place without unduly compressing the reversing springs. Thus the facing up of a shoulder on the work slightly beyond the reversing point without disturbing the reversing dogs on the sliding table can be

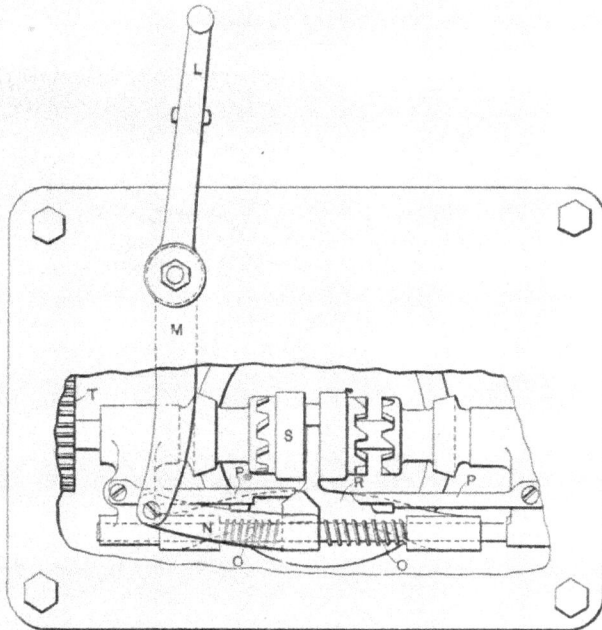

Machinery,N.Y.

Fig. 13. Reversing Mechanism of the Brown & Sharpe Grinder

readily accomplished without undue strain on any part of the reversing mechanism.

Truing the Wheel

When truing the periphery of a grinding wheel for all regular cylindrical work, a bort diamond, mounted in a suitable holder, is used, as shown in Fig. 14. This holder should be so mounted in its support that the distance from the diamond to the support *V* is as short as possible, thus avoiding spring or vibration in the holder which produces an irregular surface on the grinding wheel, appearing on the work in the form of a mottled effect or chatter. When work of large diameter is being ground, the wheel should be brought forward to the position shown in Fig. 14 when truing it off. The time consumed in moving the wheel forward with the cross-feed handwheel is more than offset by the more rapid cutting of the wheel by the diamond, and, furthermore, the surface of the wheel is in better shape, as stated.

For internal grinding, it has been proved economical by practice to true off the grinding wheel with a piece of a large wheel that has been worn down to such a small diameter as to render it useless for grinding; this piece of wheel must be harder than the wheel being trued off. This is a much quicker process than using a diamond, as the piece can be held in the hand and pressed to the grinding wheel as often as occasion requires. For this same purpose, bricks of various

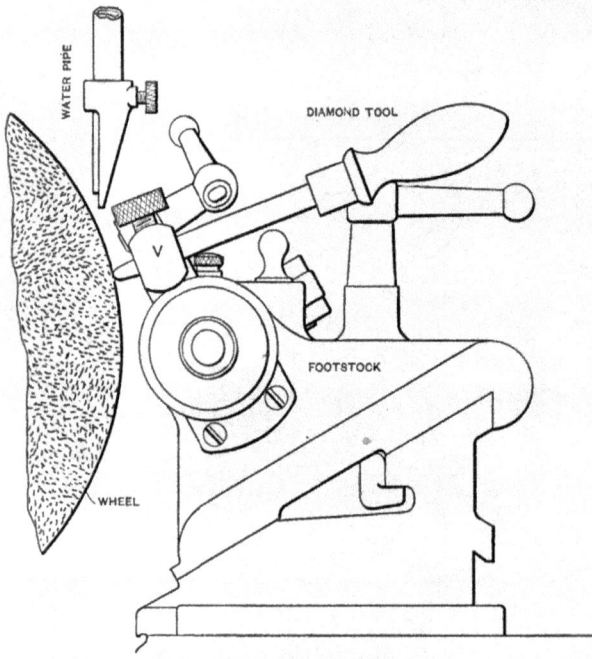

Fig. 14. The Way the Diamond Tool should be mounted for Truing the Wheel

abrasive materials have been made and are generally used when pieces of a worn-out wheel are not available.

Design of Footstock

The footstock of the grinding machine is of enough importance to warrant more attention than is generally allotted to it. The design that is in common use among grinding machine manufacturers, includes among other elements what might be called a spring-actuated spindle. The spring, which forces the center against the work, is primarily for the purpose of allowing the work to expand from the effect of the heat developed in grinding. As heat is very largely dissipated by the use of water, the more practical value of the spring-

actuated footstock is to apply a firm pressure of the center against
the work without sufficient force to distort it. This is very difficult
to do with a screw and handwheel and is accomplished on a lathe
by setting the center solid against the work, and withdrawing it
until the work can be easily turned by hand. This represents a looseness between centers intolerable on a grinder and also causes a great
waste of time. When grinding heavy work, there are two reasons
why it is often necessary to clamp the footstock spindle solid after
inserting the work in the machine. The weight of the piece of work
tends to crowd the footstock center back on account of the angle of
the center; also the momentum of the piece when the table reverses
at the footstock end of its traverse, tends to pound the center away,
and any looseness thus developed will render futile any attempt to
produce round work.

Time Required for Grinding

In Fig. 12 are shown several samples of work that have been finished on the grinding machine. Without exception every piece has
been machined complete before it is sent to the grinding department;
all threads have been cut, keyways and slots milled, holes drilled, etc.,
therefore all external and internal strains have been equalized in the
pieces before they are ground. The grinding process is the most free-
cutting process known to metal workers and should be the last cutting
process, as it distorts the work the least. The piece marked A is an
overhanging arm for the milling machine, made of machinery steel
$4\frac{1}{2}$ inches diameter and 69 inches long. These pieces require an
exceptionally good finish and are ground complete in thirty minutes
for each arm. They are revolved in the grinding machine by a pin
temporarily driven into one end near the periphery, this pin engaging
the driving arm on the headstock pulley; with this arrangement pieces
of sufficient size can be ground from one end to the other complete,
while such small pieces as those marked C and D must be turned
end for end to complete the grinding. These last pieces are about $\frac{1}{2}$
and $\frac{1}{4}$ inch diameter, respectively, and 10 inches long. They are
ground at the rate of fifteen per hour and have a limit of 0.00025 inch
either side of the dimension given. The shaft marked E is about 40
inches long and 1 inch in diameter where it is ground; this piece
can be readily dogged at one end. These are ground at the rate of
twenty minutes each with a tolerable variation of 0.00025 inch larger
or smaller. The tapered collet shown in the center of the engraving
is ground in four minutes. The milling machine spindle marked F
is ground complete in one hour. The limits are very close, *viz.*,
0.00025 inch total variation, and the taper behind the collar is ground
to a gage. The smaller spindle marked G has the same close limits
as the larger one, and is ground complete at the rate of seventy in
sixty-four hours. The spindle marked J, which is 34 inches long, has
a threaded guard over the end and on this guard the dog is clamped.
Thirty-nine of these spindles are ground complete in thirty hours.
The screw machine spindle marked H is a very difficult piece to

finish owing to the fact that it is bored out its entire length so that it is practically a hardened steel shell which is cut away at the smaller end. When grinding these spindles in large lots, they .are roughed out all over in large quantities, then finish-ground later. It requires thirty minutes to completely grind one spindle.

This last example illustrates very clearly how difficult it is to estimate the time required for grinding a piece of work, as every feature of the piece enters into the problem, and if it were not for the two slots which so cut away and weaken the small end that the grinding wheel cannot be forced into the work, these spindles could be ground in about ten minutes less time for each one. Placing the work on an arbor for grinding very seldom increases the total length of time to grind when there are several duplicate pieces in a lot, as two arbors may be employed and the operator can insert one arbor in the work while the machine is grinding the work mounted on the other arbor.

Number of Operators

The economical operation of a grinding machine presents very varied problems. It is sometimes an advantage for one man to run two machines; this is generally on long pieces where the time of actual grinding far exceeds the time of placing the work in the machines. There are, however, some jobs where two men can work very successfully on one machine. This is the case with short, large bushings that are driven on an arbor, in which case the time of changing the arbor from a finished piece to an unground piece equals or exceeds the actual time of grinding. When these conditions exist, the operator and the machine are non-productive at least half of the time, and this non-productive period can be reduced to a minimum by a helper to assist the operator.

CHAPTER IV

THE DISK GRINDER*

In any machine shop or department of a manufacturing plant where tools for manufacturing operations are made, a properly designed and equipped disk grinder should be considered almost indispensable; for a large portion of the operations most commonly done with a file, and many that are considered surface grinder, milling machine or shaper jobs, can be done better and quicker, and at less cost for files, cutters, etc., with a disk grinder.

As a simple example we will take the case of a piece of tool steel needed, say, for a box tool, a back rest, a cutter, or a forming tool, to be, say, ¼ inch thick, 1 inch wide, 2 inches long, ends and sides straight and square all around. Probably the bar steel ¼ inch by 1 inch will be enough oversize to grind on a disk to exact size, but not enough oversize to work with a milling machine, shaper or surface grinder. Even if larger stock, say 5/16 inch by 1⅛ inch, or a forging, is used,

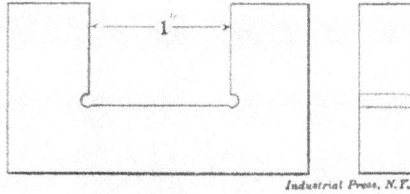

Industrial Press, N.Y.

Fig. 15. Snap Gage Finished on Disk Grinder

it is only necessary to rough one flat side and one edge down fairly close to size and finish all over on a disk grinder. For squaring the ends of one piece like this, and bringing it to exact length, the saving in time over the common way is considerable. Suppose this piece has to be hardened, and after hardening must fit a certain space. It will need truing up after hardening, and here again the disk grinder proves its adaptability.

Regarding the degree of accuracy obtainable with a disk grinder, an example may be of interest. An experienced toolmaker was with an exhibit of disk grinders at a fair. Having plenty of time on his hands, he employed a part of it in grinding up six steel pieces, each a one-inch cube. He got the pieces planed roughly in the bar, a little oversize, and sawed off a little long. In his spare time he ground them to 1-inch cubes, measuring them with a 1-inch micrometer caliper. When he had finished with them, there was no point on any of the cubes that varied more than 0.00025 inch from 1 inch. Packing them together with any combination of sides, the greatest variation

---

* MACHINERY, June, 1904.

from 6 inches as measured with a 6-inch micrometer was 0.0005 inch. All the sides of all the cubes were so nearly square with each other that no error could be detected with a hardened steel square. In grinding these cubes no fixture or clamp of any kind was used. They were laid on the swinging table, against the rib, and pressed against the wheel with the fingers.

A few examples of the application of the disk grinder to tool-room work will give a general idea of its application. Suppose a snap gage

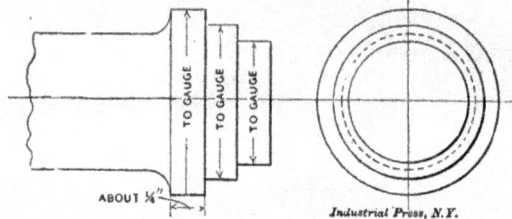

Fig. 16. Work to be Gaged

such as shown in Fig. 15 is to be made. With the disk grinder the gage can be finished all over, sides, edges and ends, and corners beveled or rounded. In hardening, the gage springs somewhat, but can easily be squared again on the disk grinder. We are now ready to grind the notch to size. Lay the piece on the swinging table, with the back edge against the rib, the wheel being in the notch. The piece is now ground on both sides without turning it over. This will make the faces of the notch parallel with each other, which they might not be if the piece were turned over. By the use of an end measure gage the snap gage in Fig. 15 is now easily completed.

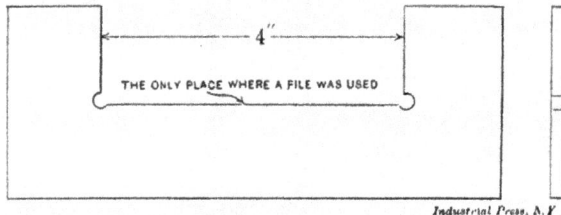

Fig. 17. Snap Gage Ground to Size on Disk Grinder

In a certain shop a job came up to be done in the turret machine. A number of cast-iron pieces, of the shape shown in Fig. 16, were to be machined. There were eight different sizes of pieces and three dimensions made to gage on each piece, making 24 dimensions in all. The largest dimension on the largest piece was about four inches. The smallest dimension on the smallest piece was about ¾ inch. A few thousand pieces of each size were to be made. Extreme accuracy was not required; a variation of 0.001 inch was allowable. A tool-maker was given the job of making a set of snap gages.

Taking the figures, he made 24 end measure pieces from 5/16-inch round drill rod, hardened them, and marked the size. He then cut the gages from ⅛-inch thick sheet steel, as shown in Fig. 17. The working faces were hardened and ground to the end measure pieces on the disk grinder, and the edges squared and the corners rounded in the same machine. The gages were not touched with a file except to smooth off the edge in the bottom of the notch.

Fig. 18. Form of Hollow Cast-iron Block used for Test

The examples given indicate the use of the disk grinder as a toolroom machine. This machine, however, is also efficient for removing large amounts of metal in a short time. The efficiency of the machine for this purpose depends largely upon the kinds of disks used. Tests were made at the shops of the Gardner Machine Co., Beloit, Wis., to determine the comparative efficiency for grinding cast iron by differ-

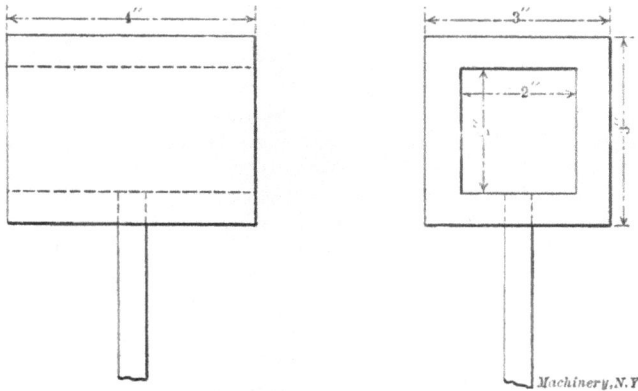

Fig. 19 Hollow Cast-iron Block used for Test

ent kinds and makes of disks, such as are commonly used in connection with disk grinders. In the following table the different kinds of disks are indicated by figures:

No. 1 indicates the Gardner improved abrasive disk No. 126. No. 5 is the regular No. 24 commercial emery cloth. No. 6 is the same in emery paper. Nos. 2, 3, and 4 are disks of excellent quality as compared to commercial emery cloth.

The disks tested were all 20 inches in diameter and all excepting Nos. 5 and 6 were No. 16 grain. The grinding was done on the ends of hollow blocks of cast iron, as shown in Figs. 18 and 19. The area ground at the end of blocks was 5 square inches. Reducing the blocks one inch in length indicated the removal of 5 cubic inches of metal. The grinding was all done on the same machine by the same operator.

The micrometer stop at the back of the table was set to grind-off a fixed amount, usually 0.050 inch, and the twelve blocks ground to the stop. The stop was then moved back 0.050 inch and the operation repeated until the blocks became too warm for efficient grinding, when

TABLE III. RESULTS OF TEST OF EFFICIENCY OF ABRASIVE DISKS

Disk Number.	Time Used in Minutes.	Stock Removed in cubic inches.	Number of Times Dressed.	Average Cutting Rate, cubic inches per min.	Cutting Rate, First Half of Time Used.	Cutting Rate, Second Half of Time Used	Life of Disk, Based on Disk No. 1.	Stock Removed, Based on Disk No. 1.
1	754	349.85	0	0.464	0.442	0.486	100. %	100.0%
2	137	42.13	6	0.307	0.344	0.270	18.1%	12.4%
3	540	113.95	0	0.211	0.238	0.184	71.6%	32.3%
4	68	27.97	2	0.411	0.546	0.276	9.0%	8.0%
5	71	2.41	4	0.034	0.062	0.006	9.4%	0.7%
6	73	12.48	2	0.171	0.273	0 069	9.7%	3.5%

they were cooled, and the time of grinding and the amount of metal removed, was noted. This was repeated until the disk was worn out or the blocks all ground up. In the latter case, new blocks were substituted and the operation continued until the disk was worn out. By reversing the blocks they were ground down until the wheel touched the handles on both sides. During this test several hundred pounds of these blocks were converted into cast-iron chips.

It will be noted in Table III that it was necessary to use a Huntington emery wheel dresser on all disks tested except Nos. 1 and 3. The dresser was used whenever the surface of the disk became dull and glazed so that it would not cut cast iron readily. The use of a dresser shortens the life of the disk, but it is absolutely necessary.

CHAPTER V

GRINDING KINKS AND EXAMPLES OF GRINDING

Grinding a Large Crankshaft[*]

A leading English chainmaker some time ago sent to the Norton Grinding Co., Worcester, Mass., a rough-turned crankshaft to be ground to the dimensions given in Fig. 20. The conditions given were that the throw must be ½ inch plus or minus 0.001 inch and that the keyway shown in Fig. 20 should line up exactly with the highest point of the eccentric. The keyway was already in the shaft when received. The following method was pursued in preparing the crankshaft for the grinder:

Two cast-iron blocks, Fig. 22, were planed to the dimensions given, and one side, E in Fig. 23, was scraped to a surface-plate. A squaring chip was then taken across a lathe face-plate and the plate was rigged with blocks and parallels as in Fig. 23. The surface E of the parallel B was also scraped to a surface-plate. When the large hole was bored, the block A, Fig. 23, was against parallel C, and when the small hole, or eccentric hole, was bored, A was moved along parallel B and block D was inserted. Tissue paper was used in both settings to insure actual contact. The large holes were bored 0.015 inch larger than the finished diameter of the crankshaft ends. After boring the small holes, a 1-inch arbor was forced into the small holes and the 60-degree center holes were turned with a lathe tool. The truth of these 60-degree holes was tested by means of a ground cone point and red lead. A tapped hole and setscrew completed each block.

The shaft was now prepared for the blocks by grinding each end a wringing fit for its block. Before doing this, the center holes in the shaft were tested and scraped to a 60-degree cone point, to insure a perfectly round shaft when ground.

The next operation was to correctly locate the keyway. For this, two blocks, A and B, Fig. 21, were made. A is a 1-inch block that tapped lightly into the keyway and projected a short distance, as shown. B is a block planed to micrometer gage, and of such a height as to bring the center line of the keyway and the center line of the crankshaft into a plane parallel to the planer surface C, Fig. 21. The proper height of B was easily found by means of micrometer measurements and deductions. Having made A and B, Fig. 21, the whole job was taken to a newly-planed planer table and the end blocks were placed on the crankshaft. A was then placed in the keyway and the crankshaft turned until A rested on B. With tissue paper under the end blocks D, Fig. 21, and between A and B, adjustments were made until all the papers held fast. The blocks D were then made secure by means

---

[*] MACHINERY, March, 1907.

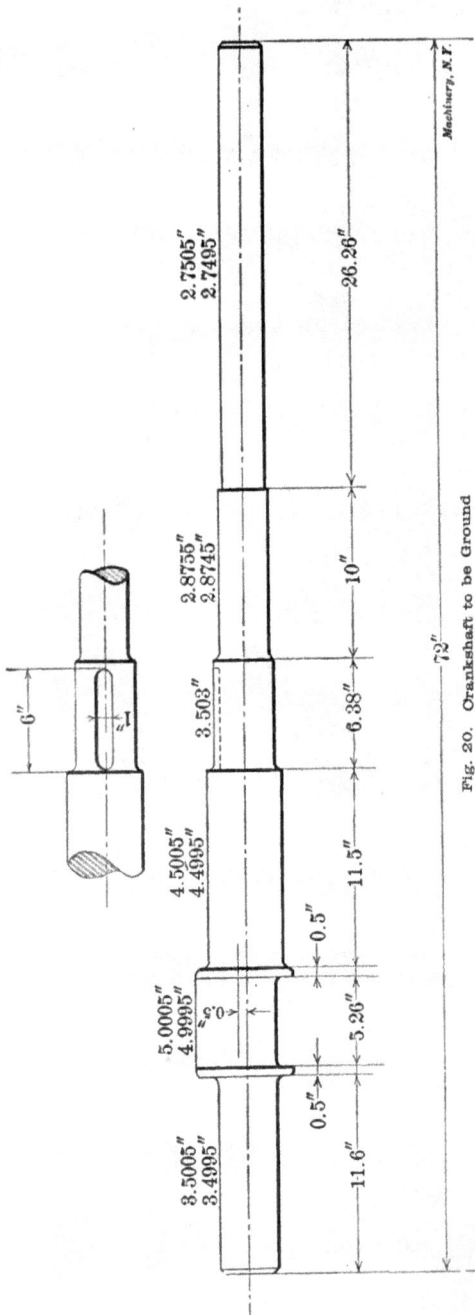

Fig. 20. Crankshaft to be Ground

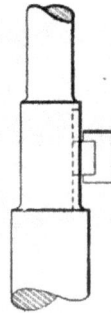

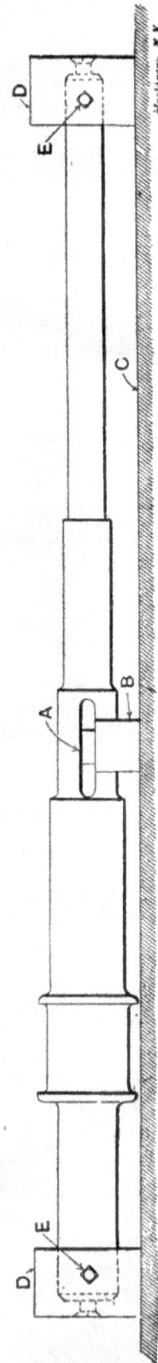

Fig. 21. Method of Mounting Crankshaft in Fixture

of the setscrews *E*. After a final test with the tissue papers, the crank-shaft was ready to have the eccentric ground. This was done on an 18-inch by 96-inch Norton plain grinder. The fillets on the eccentric were also ground at the same time.

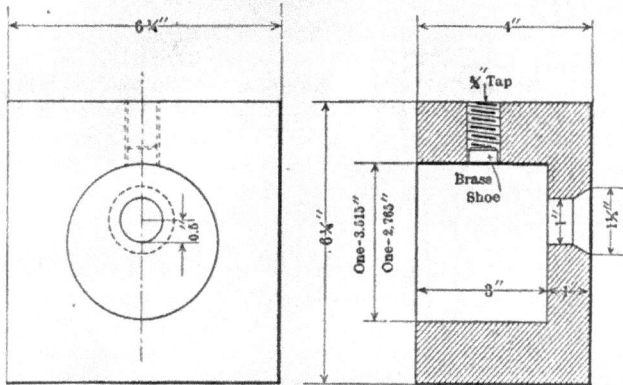

Fig. 22. Fixture for Grinding Crankshaft

The length of throw was tested in the grinder by means of a Bath indicator and a 1-inch B. & S. disk, and found to be within the required limits. When the eccentric was completed, the end blocks were re-

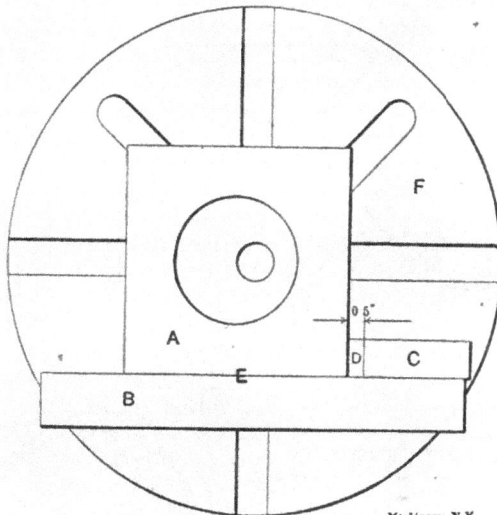

Fig. 23. Method of Boring the Fixture used for Grinding the Crankshaft

moved and the remainder of the crankshaft was ground on its own centers.

Grinding Kinks*

In the following are described some of the kinks used by toolmakers in grinding; these kinks were contributed by Paul W. Abbott.

* MACHINERY, December, 1908.

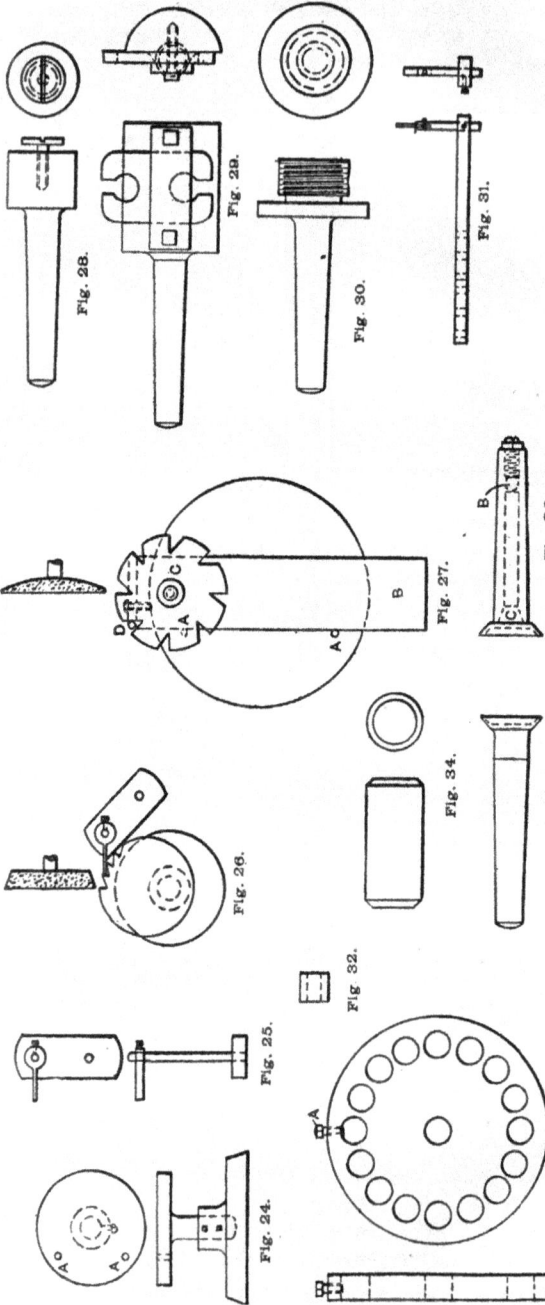

Figs. 24 to 36. Grinding Kinks

Fig. 24 is a hand grinding rest which is very handy for use on the universal grinder. It is adjustable up and down for height, and is used for hand grinding circular and straight form tools, sharpening metal slotting saws, formed cutters, etc. Fig. 26 shows the application of the hand rest to the grinding of saw teeth in a blank. The tooth rest used in connection with this operation is shown in Fig. 25. These saws are first ground on an arbor, the old teeth being ground off, leaving a perfect circle. The operator then puts on this device, setting the tooth rest so that the teeth will be about ¼ inch apart, and grinds around by hand, not quite bringing each tooth to a sharp point. On the last nine or ten teeth he evens up any inaccuracy in the spacing, the wheel being trued off to the exact shape of tooth space wanted.

Fig. 27 shows a device for accurately sharpening formed cutters up to 3 inches diameter, which is used when the cutter grinder has another job in it, or could be used to advantage where there was no surface or cutter grinder. The device consists of the cast-iron slide *B*, at the end of which is a tapped hole *C*, with a small fillister head screw which holds the various sizes of bushings which fit the holes in the cutters. On the same end is the index pin *D*, which is adjustable back and forth. In operation, the hand rest shown in Fig. 24 is also used, and the pins *A* are lined up parallel with the forward travel of the wheel, and so that the cutting face of the wheel is on a line with the center of the bushing. The cutter is then slipped on over the bushing and the index pin is set so that the required amount will be ground from the face of the tooth. The operator brings the wheel up to the proper position and then pushes the slide forward until the wheel has reached the bottom of the tooth space; he then withdraws the slide and indexes to the next tooth, and so on, tooth after tooth. It will be noticed that the index pin rests against the back of the tooth, which means that upon the previous milling of the teeth depends the accuracy of the grinding; but on the standard cutters furnished by numerous concerns this spacing will be found accurate enough.

Fig. 28 is a center for the head-stock for holding small forming tools of odd size, or threaded pieces which are to be ground on the periphery. The tools are simply clamped to the face of the center, and trued up by an indicator. Fig. 29 is a device for the tool grinder for grinding snap gages, where there is no surface grinder for this class of work. The shank of this device is made to fit the head-stock, and the gages are clamped to it by a small strap and two screws. This fixture revolves while in use, and the jaws of the gage are ground by feeding a thin wheel in and out by hand. Revolving the device insures perfectly straight gage faces. Fig. 30 shows a center for the universal grinder for holding a standard line of large end milling cutters with threaded holes, while sharpening. The head-stock is swung around at right angles to the ways, and with a long support for the tooth rest (Fig. 31), which is bolted to the platen, the cutters are ground very handily by throwing in the feed and grinding one tooth, and then, before the wheel comes back, indexing to the next tooth, and so on.

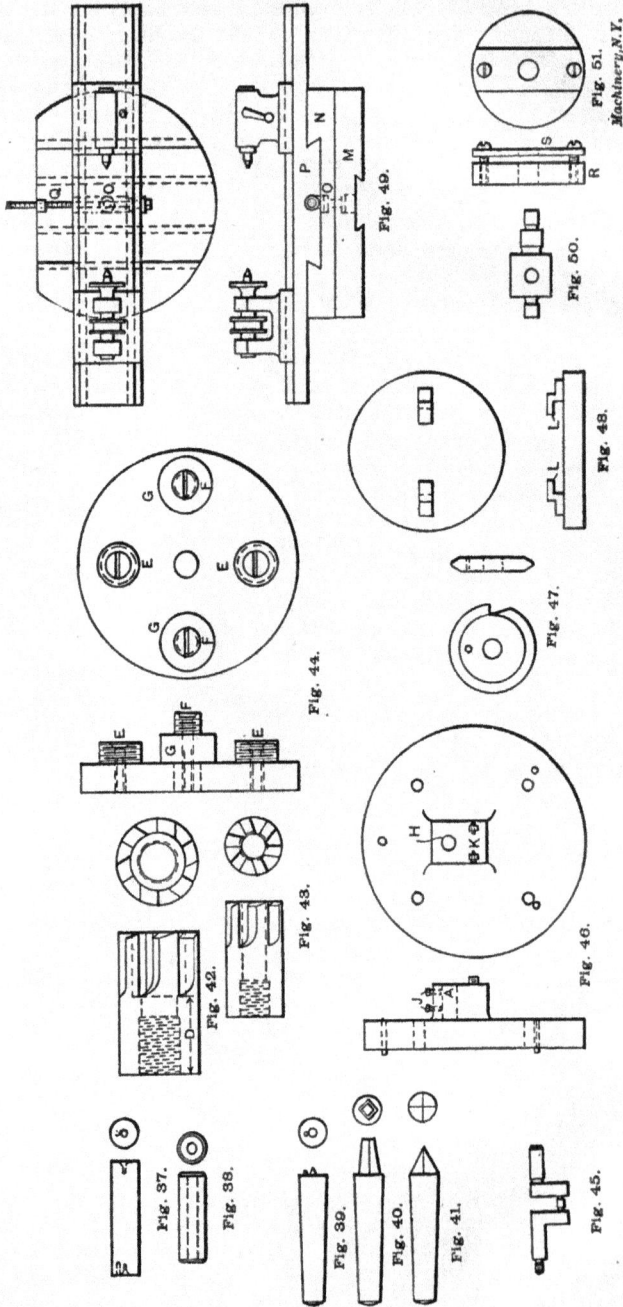

Figs. 37 to 51. Grinding Kinks

Fig. 32 shows a hardened roller which is ground all over, and Fig. 33 the fixture for the universal grinder for grinding the sides of this roll. This plate was made of cast iron, with both sides ground and with each hole ground to 0.0005 inch over standard size. Each hole has a ¼-inch set-screw, as shown at A. In operation, the plate is fastened to the face-plate by a draw-back rod, and the head-stock is swung around at right angles. As the plate revolves, 16 rolls are ground at once, first on one side, and then the plate is turned and the other side ground, the rolls being made to standard length by using a depth gage. The hardened roll shown in Fig. 34, which is used on swaging machines, is held by the centers shown in Fig. 35 and 36, when being ground. Fig. 35 is the head-stock center cupped out on the end to fit the beveled end of the roll. This center drives the roll by friction, the pressure being obtained by the spring tail-stock. Fig. 36 is the tail-center, which is in two parts, the inner spindle running with the roll and being adjusted by the screw in the end so that the thrust is taken by the ball B, the tapered portions C just clearing each other. Other methods of grinding rolls are shown in Figs. 37 to 41. One example of grinding is shown in Fig. 37, and its center in Fig. 39. The roll is driven by a pin on the center, which engages with a corresponding hole in the work. A better method is to center the roll and then in one end drive a square 60-degree punch, using the square center shown in Fig. 41 for driving the work while grinding. Another good method for hollow rolls, such as shown in Fig. 38, is to use a 15-degree square center, such as shown in Fig. 40, the end of which just enters the hole.

Figs. 42 and 43 show two end mills. The smaller one is fastened inside of the larger when in use, and when in position rests against the bottom of the hole and projects outside a definite distance. The length D is standard in all these mills. Fig. 44 shows the fixture for grinding two pairs of these mills at a time, so that the same amount will be taken off of both the short and long ones. Threaded bushings E fit the larger size mills, and F, the smaller. The collars G are of such thickness that the cutting face of the smaller mill is brought into the same plane as the larger, and so when grinding an equal amount is removed from the face of each mill. The plate is held to the face-plate by a draw-back rod. The head-stock is swung at right angles, and with the fixture revolving, the wheel traverses back and forth across the faces of the mills. The mills are then taken to a cutter grinder and backed off.

Fig. 45 shows a small crankshaft, and Fig. 46 the fixture for grinding the pin. The bearings are first ground on centers in the usual way. The fixture is of cast iron and is held to the face-plate by screws and dowel pins. In the making of this fixture the hole H was ground out to the size of the bearing, and then the fixture was correctly located and doweled to the regular face-plate. The crank, while being ground, is held by the set-screws J and the screws K, which are set against the crank on either side.

The grinding of formed cutters, similar to the one shown in Fig. 47, so that they will be interchangeable, is very interesting. The error limit is 0.00025 inch. The grinder used is a Norton universal tool and cutter grinder. After hardening, the cutters are first ground to a definite thickness. For this operation they are held against the face-plate by a draw-back chuck. The next operation is grinding the beveled sides, which is accomplished by holding the cutters against a small face-plate by a draw-back chuck. The correct angle of bevel is obtained with the protractor, and to get the correct diameter of the bevel sides, and to insure that the bevel sides stand exactly in the same relation to each other, the gage shown in Fig. 48 is used. This gage is hardened and ground all over, and the two gaging points L are set a predetermined distance apart and as near the same height from the platen as mechanical means can make them. It is obvious that cutters which are all ground the same thickness, and which will pass through this gage with the beveled sides both touching the gage points with equal pressure, will interchange within pretty close limits. The operator grinds one bevel side at a time, trying the work every little while in this gage; when one side passes through the gage the cutter is turned around and the other bevel ground. For grinding the radius on the periphery and bringing the cutter to the correct diameter, the radius grinding fixture shown in Fig. 49 is used. The dovetailed base M is fitted to the platen of the grinder, and upon this base is a sliding base N which is pivoted to M by a bolt O. Upon the base N there is an auxiliary platen P which can be adjusted back and forth by the screw Q for getting the proper radius. This auxiliary platen is made the same as the machine platen so that the regular head- and tail-stocks will go on it. A cutter is placed on a special arbor and the platen P adjusted to give the correct radius. The wheel is then brought up and the cutter is ground to the correct diameter, the curved face being obtained by swinging the base N back and forth by hand in an arc of a circle, with bolt O as a center.

Another ingenious scheme is shown in Fig. 51. Three or four pieces similar to the one shown in Fig. 50 were to have the holes ground out. With an independent 4-jawed chuck this would have been easy, but there was no such chuck; and as there would never be any more of these pieces to be ground the fixture for doing the work had to be inexpensive. The face-plate could not be used, as the pieces were smaller than the hole in the face-plate. The operator thought awhile, and then hunted around a few minutes and found a large washer R, tapped two holes in it, filed up the sheet steel strap S, and with a couple of machine screws was ready to begin. The washer was first put in the universal chuck and the outer side ground. One of the pieces was then clamped in place, and after putting on the internal grinding attachment it was ready to be ground.

Selection of Wheel*

The following little hints regarding grinding, taken from a booklet issued by the Norton Co., Worcester, Mass., will prove of value to all who have to do with grinding machines and grinding.

---

* MACHINERY, August, 1908.

Don't believe that all materials can be ground equally well with one and the same wheel.

Get the proper wheel for the work.

You would not expect to turn all kinds of lathe work with one tool having only one form of cutting edge. The grinding wheel is a tool for cutting.

Different shapes of work, different kinds of metal, require different cutting edges when grinding as well as when turning. Different grades and grains of wheels are required for different kinds of work.

Grinding wheels are numbered from coarse to fine, and graded from soft to hard. The grade is denoted by the letters of the alphabet from E to Z.

Don't decide on the wheel without knowing the work.

Spindle speed and character of the material, shape of work to be ground, and surface of wheel in contact are prime factors.

In cylindrical grinding, speed of work, diameter of work and depth of cut must all be reckoned with in the selection of the right combination of grain and grade.

The condition of the machine affects the efficiency of the wheel. Heavy machines with large wheel spindles and massive wheel support call for a wheel different from those for lighter machines with smaller spindles.

Don't order a certain grade of wheel merely because that grade is used on similar work in another plant.

Don't use a hard wheel to economize—it is production you are after.

A hard wheel is more likely to change the temperature of the work or to become glazed than a soft one; furthermore, it requires more power to do the same amount of work.

It is a common error to assume that a wheel for grinding steel and cast iron, chilled iron and hardened steel must be as fine as the surface desired. A coarse wheel will produce a fine finish if the proper relations between grade, depth of cut, speed of work, speed of wheel, etc., are observed.

When grinding brass and the softer bronzes, the wheel must be as fine as the finish required. Bronzes with "manganese" or "phosphor" permit the use of coarser wheels.

Don't get a wheel made for soft steel for use on hard steel.

For a fine finish on hard stock, a coarse wheel may be necessary, and the harder the stock, the coarser the wheel.

When ordering wheels, don't forget the diameter, width, style of face, arbor holes, description of work, speed of spindle, and the number and letter denoting the combination of grain and grade, if known.

The width of the wheel should be in proportion to the amount of the material to be removed with each revolution of the work.

If you reduce the width of the wheel you must use a finer feed, and consequently do less work.

Mounting

Never mount wheels without flanges.

Flanges should be at least one-third the diameter of the wheel; one-half is recommended. Flanges should be concave—never straight or convex.

Use fiber or rubber washers a trifle larger than the diameter of the flanges, or flanges with soft metal facings.

Hooded machines are desirable when practicable.

Truing

Don't start work on a new wheel until you are sure it runs true.

Always have a wheel dresser handy for truing wheels for off-hand grinding.

Never use a dresser on wheels that grind circular work on centers.

For truing wheels used on plain cylindrical and universal grinding machines, cutter and reamer grinders, etc., the diamond is recommended. To obtain the best results it is absolutely necessary.

Never attempt to true a wheel for circular grinding unless the diamond is held in a rigid toolpost on the table of the machine. You cannot do good work with such a wheel when it is trued "by hand."

To get a truly ground surface you must keep the face of the wheel true.

Speed

Don't start grinding until you know the speed is right—not "near enough," but right.

Even a slight variation in speed may be the cause of success or failure of any wheel.

Failure is sometimes turned into success by merely changing the speed of either the wheel or work.

Speed up the spindle as the diameter of the wheel is decreased. Approximately the same peripheral rate should be maintained as the wheel wears down.

Complaint is sometimes made that wheels appear to be softer toward the center. Usually this is because the same surface rate of speed is not maintained as the wheel is reduced in diameter. This causes the wheel to wear away faster and appear softer. It is also true that while the grade of the wheel may be uniform throughout, yet the smaller line of contact due to the smaller diameter will cause the wheel to appear softer.

Increasing the speed of a grinding wheel gives the effect of a harder wheel; decreasing the speed gives the effect of a softer wheel.

For surface grinding it is customary to run wheels at a somewhat slower rate of speed than for general grinding. A speed of 4,000 to 5,000 surface feet is usually employed.

Wheels are run in actual practice from 4,000 to 6,000 feet per minute.

General Suggestions

Transferring a wheel worn down to a small diameter from a large machine to a small one is good practice.

Keep the tickets or tags which are sent on the wheels in a record book, so that if a wheel is not satisfactory, reference can be made to order number when making complaint. It is equally valuable as a reference when ordering duplicate wheels.

Don't use the wrong wheel on a job because it will require a few minutes' time to change wheels. A stop-watch will prove to you that changing wheels is cheaper.

There is seldom a case where one and the same wheel can be used on all work without a greater loss of time than the change of wheel would involve. Many times the time saved in grinding a single piece more than pays for changing the wheel.

Considerable difference in diameters of work will affect the cutting quality of a wheel on any given material.

A successful wheel on the small diameters may work much slower on the larger diameters.

The wheel most suitable for work of very large diameter may wear away too fast on work of smaller diameter.

A suitable wheel for small diameters may cause chatter on pieces of large diameters.

Don't grind circular work dry.

A good wheel will grind in water, soda water or oil.

Water keeps the wheel working cool and increases grinding production.

Soda water keeps the work and the machine from rusting.

Oil in soda water increases the wheel's effectiveness.

The particles from a grinding wheel do not adhere to steel. Don't let any one convince you to the contrary.

Grinding is profitable for removing stock as well as for finishing.

Keep the face of the wheel true and parallel with axis of spindle.

Vibration makes grinding wheels wear.

Keep all rests adjusted close to the wheel, otherwise work is liable to be caught and injury result.

Keep boxes well oiled and adjusted.

When practicable, indicate on each machine the revolutions of spindle and size of wheel to be run upon it.

Don't disregard the setting up instructions that go with the grinding machine.

CHAPTER VI

COST OF GRINDING*

To figure, with any degree of accuracy, the cost of commercial wet grinding, requires considerable experience in the use and management of the machine, in order to be as closely approximated as lathe work. There seems to be a great difference in operators, due partly, no doubt, to the fact that the grinder has not yet become as generally used as other standard machine tools. A great many operators seem to be afraid to push their machines, and spend a good deal of time in useless calipering. They seem to forget that if they have several thousandths of an inch to take off a piece and are feeding in one or two thousandths of an inch at each reversal of the machine, they need not caliper until within one or two thousandths of an inch of size. Another class seems to think that because grinding is a finishing job, it must be nursed.

As a matter of fact, there is no machine which so rapidly and accurately responds to the touches of an operator as the wet grinding machine. Of course, there are delicate pieces and certain shapes which

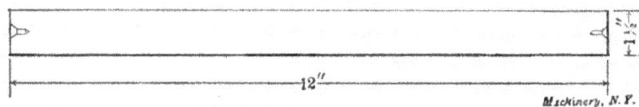

Fig. 52. Plain Cylindrical Piece to be Ground

have to be carefully handled, but the usual run of work is so simple that any good apprentice can be put on it and taught in a short time.

The work usually comes from the lathe with approximately 1/64 to 1/32 inch stock to be removed. The work is then completed by a few reversals of the grinding machine with a feed nearly the full width of the wheel, and a cut of two to four thousandths of an inch until nearly up to size, and then a much slower traverse per revolution for finishing, according to the kind of finish desired. To obtain the best speed, the limits required on the lathe must not be made too narrow, from 1/64 to 1/32 inch being admissible for ordinary work, and more on large work; for the facility of the grinder in finishing work is far in excess of the lathe, and the latter must be relieved of all the finishing possible.

To figure the actual time for removing stock on the grinder, we must take into account the longitudinal traverse of the wheel for each revolution of the work, the surface speed of the work and the depth of the cut. The latter must be varied according to the nature of the material, greater or less according to whether it is hard or soft; and the traverse per revolution of work is lessened if a fine finish is desired. The shape of the piece also somewhat affects both of these

---

* MACHINERY, October, 1906.

points, as long, thin pieces require a slower traverse and lighter cuts.

Take, for instance, the plain piece, Fig. 52;
material, hardened steel. For this a work surface speed of 15 feet, or about 37 revolutions per
minute would be suitable. Assuming we have
a wheel 18 inches in diameter and 1½ inch face,
a traverse of two-thirds the face of the wheel or
one inch per revolution of work is usual. This
would require 12 revolutions to pass the length of
the piece, plus 1 revolution for clearance, or for
dwell if there happens to be a shoulder. This
would make, roughly, three reversals a minute.

On a medium-sized machine, an automatic feed
equivalent to a work reduction of about 0.002
inch would be suitable, or a reduction of about
0.006 inch per minute. If the work came with
an average allowance of 0.030 inch for grinding,
it would require theoretically 5 minutes' actual
grinding time to rough this piece down. To this
must be added the time for handling the work,
adjusting the machine and back-rests (in this
case only one rest would be used), calipering the
work and finishing. This time will amount to as
much as the grinding time with most operators
(most of it being taken up in finishing), which
would make the actual time about ten minutes
apiece. As a matter of fact, work of this size is
actually being ground at the rate of seven or eight
pieces per hour.

If a fine finish were desired, a higher work
speed and slower traverse would be required. For
a very fine finish a work speed of 45 feet surface
speed and traverse of 1/6 inch per revolution would
be suitable for finishing, with, of course, a very
much smaller feed. This change in the work and
traverse speed could be made when the work is
nearly up to size, and would probably require
about three minutes. If the piece were of soft
steel, a deeper cut could be taken and a wider
traverse, a cut of 0.003 inch and a traverse nearly
up to the width of the wheel being admissible. In
grinding long shafts it is necessary to allow pro-
portionately more time for adjusting back rests and
for calipering, to insure that the piece be straight.
This often takes twice the actual grinding time.

Now let us look at the more complicated piece,
Fig. 53. This will have to be done on a larger machine, and the larger
machines are slower to handle. This piece is a piston rod of 40 carbon

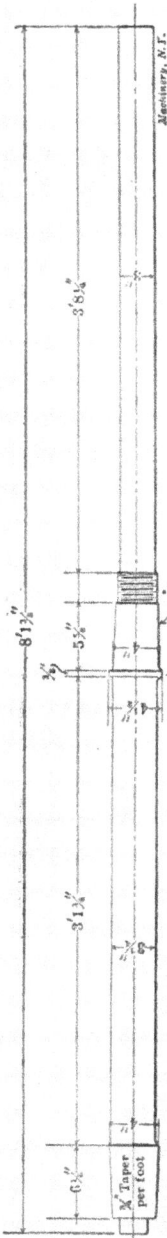

Fig. 53. A more Complicated Piece which is to be Ground

steel. We will use for this a 20-inch wheel of 2½ inches face. A suitable traverse for this wou!d be 2 inches per revolution and a surface speed of 15 feet would make about 19 revolutions for the part 3 inches in diameter, and about 15 for the part 3¾ inches. The figures would be about as follows:

Total amount to be removed, 0.060 inch; amount per reversal, 0.004 inch; number of reversals required, 15.

3 inches diameter, to cross once, 1 1/5 minute; total for
    15 reversals .... 18 minutes
3¾ inches diameter, to cross once, 1½ minute; total for
    15 reversals .... 22 minutes
Tapers, both, to cross once, 2/5 minute; total for 15 re-
    versals .... 6 minutes
Setting up and adjusting............................. 10 minutes
                                  —
    Total .... 56 minutes

If it be desired to put a radius on the wheel and grind the fillets at shoulder *A*, about 10 minutes more shou!d be allowed; and if there were more than one piece to be done, considerable time could be saved in setting for the tapers.

CHAPTER VII

THE BURSTING OF EMERY WHEELS[*]

In 1902 some important tests of the strength of emery wheels were undertaken at the Case School of Applied Science, Cleveland, Ohio, under the direction of Prof. H. Benjamin. Fifteen wheels of various makes were tested to destruction. The results of these tests are given in the following.

Most manufacturers of this class of wheels test them for their own information, but the results are not generally given to the public. At the Norton Emery Wheel Works all wheels are tested before leaving the shop at a speed double that allowed in regular service, and occasionally wheels are burst to determine the actual factor of safety.

Emery-wheel accidents are not uncommon, but can usually be traced to the carelessness of the operator. One common cause of failure is allowing a small piece of work to slip or roll between the wheel and the rest.

The wheels selected for the experiments were all of the same size, being sixteen inches in diameter by one inch thick, and having a hole one and one-half inch in diameter. The object of the experiment being to determine the bursting speed of such wheels as are actually on the market, emery wheels were obtained through various outside

---

[*] MACHINERY, July, 1903.

parties without indicating to the agents or manufacturers the use to be made of them. In this way wheels of six different makes were obtained, the label on each wheel showing usually the maker's name, the grade number or letter, the quality of emery, and the speed recommended for use. As shown in Table IV, giving the results, the working speed varied in the different wheels from 1,150 to 1,400 revolutions per minute, the average being about 1,200 revolutions per minute. For a diameter of sixteen inches this corresponds to a peripheral velocity of about 5,000 feet per minute. The table also shows that the fineness of the emery varied from ten to sixty, the average being about thirty.

The wheels were held between two collars, each six and one-eighth inches in diameter and concaved, so as to bear only on a ring three-fourths of an inch wide at the outer circumference.

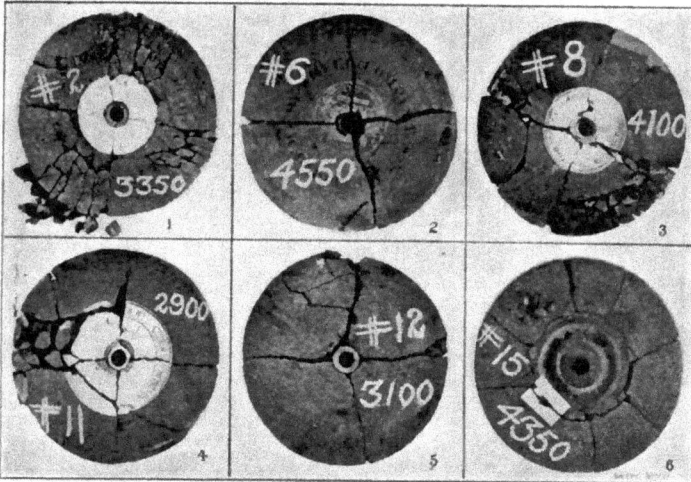

Fig. 54. Various Ways in which Emery Wheels Burst

Table IV shows the results of the experiments in detail, and needs but little explanation. The illustrations in Fig. 54 show characteristic fractures, and the appearance of various wheels after bursting. Wheels numbered 1, 2, and 3 in the table were of one make, and showed a remarkable uniformity in strength. Nos. 4, 5, 8, and 9 were all made by one firm; the two latter wheels were of finer grain than the others, and showed a correspondingly greater strength. Nos. 6 and 7 contained a layer of brass wire netting imbedded in the emery, and were about one-third stronger than the average of the ordinary wheels. The wheels numbered 10 and 11 were the weakest among those tested, but had an apparent factor of safety of between five and six. Nos. 12 and 13, of still another make, burst at about the average speed. Wheels Nos. 14 and 15 were so-called vulcanized wheels, containing rubber in the bond, and intended for particularly severe service. These showed, as was expected, rather more than the average strength.

An examination of the last two columns in the table shows that the wheels burst at speeds varying from two and one-quarter to three and three-quarters of the working speed, and accordingly had factors of safety varying from five to thirteen.

It is, then, apparent that any of these wheels were safe at the speed recommended, ánd would not have burst under ordinary conditions. At the same time, considering the violent nature of the service and the shocks to which they are exposed, it would seem that the factor

TABLE IV. RESULTS OF TEST ON EMERY WHEELS

No of Test.	Grade Mark.	No. of Emery.	WORKING SPEED.		BURSTING SPEED.		Speed Ratio	Factor of Safety.
			Revs. per Minute.	Feet per Minute.	Revs per Minute.	Feet per Minute.		
1	4 5	20	1,200	5,030	3,100	13,000	2.58	6.67
2	4.5	20	1,200	5,030	3,200	13,400	2.67	7.14
3	4.5	20	1,200	5,030	3,350	14,020	2.79	7.73
4	Q	30	1,250	5,230	3,750	15,700	3 00	9.00
5	Q	30	1,250	5,230	2 750	11,500	2.20	4.84
6	H	30	1,400	5,870	4 550	19,050	3.25	10.56
7	H	30	1,400	5,870	4,600	19,200	3.28	10 76
8	O	36	1,250	5,230	4,100	17,200	3.28	10.76
9	O	36	1,250	5,230	4,125	17,250	3.30	10.89
10	2 5	60	1,150	4,830	2,750	11,500	2.39	5.71
11	2.5	60	1,150	4,830	2,900	12,100	2.52	6.35
12	M. H.	14	1,200	5,030	3.100	12,970	2.58	6.66
13		24	1,200	5,030	3,800	15,900	3.17	10.00
14	H	10–12	1,200	5,030	4.100	17,200	3.42	11.70
15	H	10–12	1,200	5,030	4,350	18,200	3.62	13.10

Tests 6 and 7; wheels made with wire netting; tests 14 and 15, with vulcanized rubber.

of safety for emery wheels should be large. In comparison with those generally used in machines, a factor of eight or ten would seem small enough. It may also be said that such a variation in strength between wheels of the same make and grade, as for instance, that between Nos. 4 and 5, indicates a lack of uniformity which causes distrust. The fractures were in the main radial, as may be seen from Fig. 54, the wheel splitting in three, four or five sectors as might chance. It may be assumed that these radial cracks started from the rim, where the velocity and stress were greatest, but it is a fact worthy of notice that in nearly every instance the cracks radiated from points where the lead bushing projected into the body of the wheel.

www.ingramcontent.com/pod-product-compliance
Lightning Source LLC
Chambersburg PA
CBHW080722220326
41520CB00056B/7370